Boolean Algebra in Digital Electronics

BY PRASUN BARUA

1

ABOUT

"Boolean Algebra in Digital Electronics" is a comprehensive and indispensable guide for both novice and experienced enthusiasts delving into the world of digital electronics. This book explores the fundamental concepts of Boolean Algebra, offering readers a thorough understanding of the core principles that underpin modern digital technology. Beginning with a meticulous introduction to Boolean Algebra, the book elucidates the foundational concepts, such as Boolean functions, Boolean laws, and rules. It systematically unravels the intricate web of logic gates, including the AND Gate, OR Gate, NOT Gate, NAND Gate, NOR Gate, XOR Gate, and XNOR Gate, demystifying their operations and applications. With clarity and precision, this book equips readers with the knowledge required to manipulate binary information and construct logical circuits effectively. Whether you are a student seeking to grasp the essentials of digital electronics or a seasoned engineer looking to reinforce your expertise, "Boolean Algebra in Digital Electronics" serves as an invaluable resource, offering a comprehensive exploration of the key concepts and tools that power the modern digital world.

TABE OF CONTENTS

CHAPTER   **PAGE**

Chapter 1: Boolean Algebra   4

Chapter 2: Boolean Functions   8

Chapter 3: Boolean Laws & Rules   11

Chapter 4: Logic Gates   21

Chapter 5: AND Gate   27

Chapter 6: OR Gate   31

Chapter 7: NOT Gate   35

Chapter 8: NAND Gate   37

Chapter 9: NOR Gate   43

Chapter 10: XOR Gate   49

Chapter 11: XNOR Gate   53

CHAPTER 1: BOOLEAN ALGEBRA

The logical symbol 0 and 1 are used for representing the digital input or output. The symbols "1" and "0" can also be used for a permanently open and closed digital circuit. The digital circuit can be made up of several logic gates. To perform the logical operation with minimum logic gates, a set of rules were invented, known as the **Laws of Boolean Algebra**. These rules are used to reduce the number of logic gates for performing logic operations.

The Boolean algebra is mainly used for simplifying and analyzing the complex Boolean expression. It is also known as **Binary algebra** because we only use binary numbers in this. **George Boole** developed the binary algebra in **1854**.

Rules in Boolean algebra

1. Only two values(1 for high and 0 for low) are possible for the variable used in Boolean algebra.
2. The overbar(-) is used for representing the complement variable. So, the complement of variable C is represented as .
3. The plus(+) operator is used to represent the ORing of the variables.
4. The dot(.) operator is used to represent the ANDing of the variables.

Properties of Boolean algebra

These are the following properties of Boolean algebra:

Annulment Law

When the variable is AND with 0, it will give the result 0, and when the variable is OR with 1, it will give the result 1, i.e.,

$B.0 = 0$

$B+1 = 1$

Identity Law

When the variable is AND with 1 and OR with 0, the variable remains the same, i.e.,

B.1 = B

B+0 = B

Idempotent Law

When the variable is AND and OR with itself, the variable remains same or unchanged, i.e.,

B.B = B

B+B = B

Complement Law

When the variable is AND and OR with its complement, it will give the result 0 and 1 respectively.

B.B' = 0

B+B' = 1

Double Negation Law

This law states that, when the variable comes with two negations, the symbol gets removed and the original variable is obtained.

((A)')' = A

Commutative Law

This law states that no matter in which order we use the variables. It means that the order of variables doesn't matter in this law.

A.B = B.A

A+B = B+A

Associative Law

This law states that the operation can be performed in any order when the variables priority is of same as '*' and '/'.

(A.B).C = A.(B.C)

(A+B)+C = A+(B+C)

Distributive Law

This law allows us to open up of brackets. Simply, we can open the brackets in the Boolean expressions.

A+(B.C) = (A+B).(A+C

A.(B+C) = (A.B)+(A.C)

Absorption Law

This law allows us for absorbing the similar variables.

B+(B.A) = B

B.(B+A) = B

De Morgan Law

The operation of an OR and AND logic circuit will remain same if we invert all the inputs, change operators from AND to OR and OR to AND, and invert the output.

$(A.B)' = A'+B'$

$(A+B)' = A'.B'$

CHAPTER 2: BOOLEAN FUNCTIONS

The binary variables and logic operations are used in Boolean algebra. The algebraic expression is known as **Boolean Expression**, is used to describe the **Boolean Function**. The Boolean expression consists of the constant value 1 and 0, logical operation symbols, and binary variables.

Example 1: F=xy' z+p

We defined the Boolean function **F=xy' z+p** in terms of four binary variables x, y, z, and p. This function will be equal to 1 when x=1, y=0, z=1 or z=1.

Example 2:

$$F(A, B, C, D) \quad = \quad A + \overline{BC} + ACD \qquad Equation\ No.1$$
$$Boolean\ Function \quad Boolean\ Expression$$

The output Y is represented on the left side of the equation. So,

$$Y = A + BC + ACD$$

Apart from the algebraic expression, the Boolean function can also be described in terms of the truth table. We can represent a function using multiple algebraic expressions. They are their logically equivalents. But for every function, we have only one unique truth table.

In truth table representation, we represent all the possible combinations of inputs and their result. We can convert the switching equations into truth tables.

Example: F(A,B,C,D)=A+BC'+D

The output will be high when A=1 or BC'=1 or D=1 or all are set to 1. The truth table of the above example is given below. The 2^n is the number of rows in the truth table. The n defines the number of input variables. So the possible input combinations are $2^3=8$.

Inputs				Output
A	B	C	D	F
0	0	0	0	0
0	0	0	1	1
0	0	1	0	0
0	0	1	1	1
0	1	0	0	1
0	1	0	1	1
0	1	1	0	0
0	1	1	1	1
1	0	0	0	1
1	0	0	1	1
1	0	1	0	1
1	0	1	1	1
1	1	0	0	1
1	1	0	1	1
1	1	1	0	1
1	1	1	1	1

Methods of simplifying the Boolean function

There are two methods which are used for simplifying Boolean function. These functions are as follows:

Karnaugh-map or K-map

De-Morgan's law is very helpful for manipulating logical expressions. The logic gates can also realize the logical expression. The k-map method is used to reduce the logic gates for a minimum possible value required for the realization of a logical expression. The K-map method will be done in two different ways, which we will discuss later in the **Simplification of Boolean expression** section.

NAND gates realization

Apart from the K-map, we can also use the NAND gate for simplifying the Boolean functions. Let's see an example:

Example 1: F(A,B,C,D)=A' C'+ABCD'+B' C' D+BCD'+A'B'

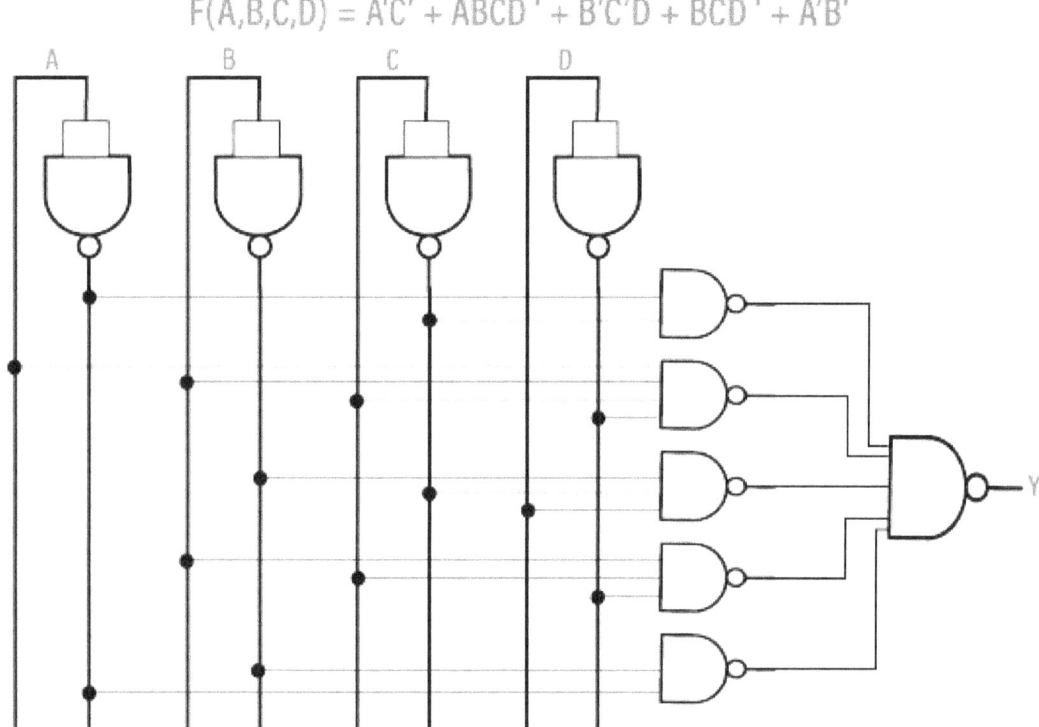

CHAPTER 3: LAWS AND RULES OF BOOLEAN ALGEBRA

In simplification of the Boolean expression, the laws and rules of the Boolean algebra play an important role. Before understanding these laws and rules of Boolean algebra, understand the Boolean operations addition and multiplication concept.

Boolean Addition

The addition operation of Boolean algebra is similar to the OR operation. In digital circuits, the OR operation is used to calculate the sum term, without using AND operation. A + B, A + B', A + B + C', and A' + B + + D' are some of the examples of 'sum term'. The value of the sum term is true when one or more than one literals are true and false when all the literals are false.

Boolean Multiplication

The multiplication operation of Boolean algebra is similar to the AND operation. In digital circuits, the AND operation calculates the product, without using OR operation. AB, AB, ABC, and ABCD are some of the examples of the product term. The value of the product term is true when all the literals are true and false when any one of the literal is false.

Laws of Boolean algebra

There are the following laws of Boolean algebra:

Commutative Law

This law states that no matter in which order we use the variables. It means that the order of variables doesn't matter. In Boolean algebra, the OR and the addition operations are similar. In the below diagram, the OR gate display that the order of the input variables does not matter at all.

For two variables, the commutative law of addition is written as:

A+B = B+A

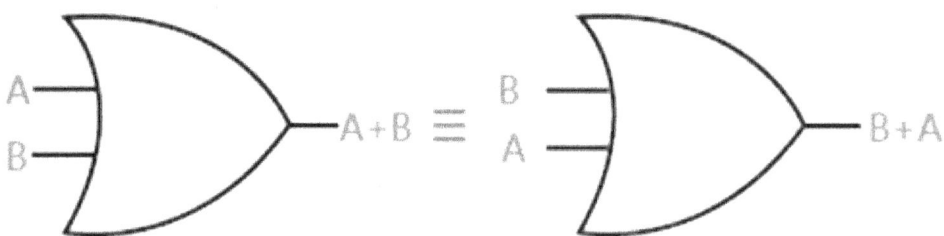

For two variables, the commutative law of multiplication is written as:

A.B = B.A

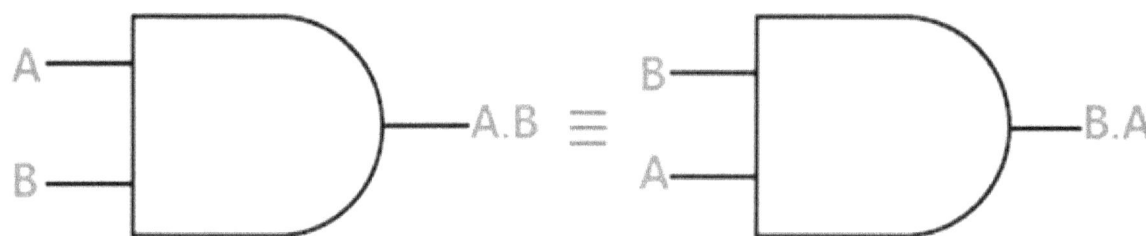

Associative Law

This law states that the operation can be performed in any order when the variables priority is same. As '*' and '/' have same priority. In the below diagram, the associative law is applied to the 2-input OR gate.

For three variables, the associative law of addition is written as:

A + (B + C) = (A + B) + C

Associative Laws

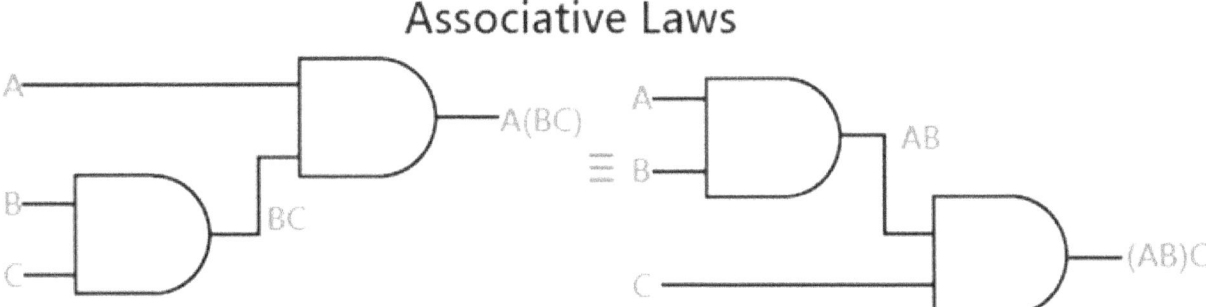

For three variables, the associative law of multiplication is written as:

A(BC) = (AB)C

According to this law, no matter in what order the variables are grouped when ANDing more than two variables. In the below diagram, the associative law is applied to 2-input AND gate.

Associative Laws

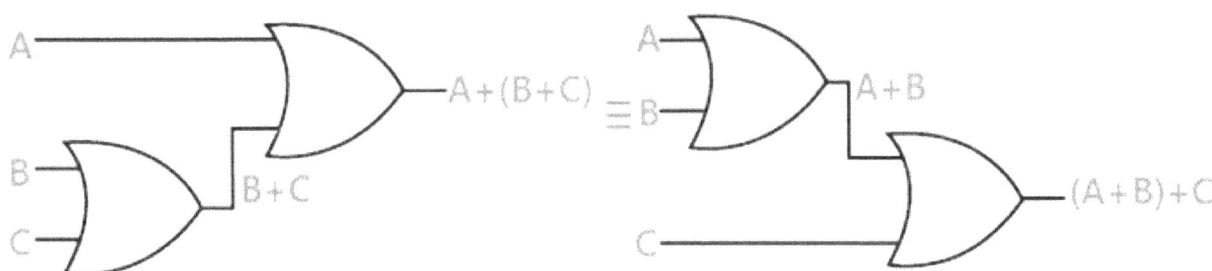

Distributive Law:

According to this law, if we perform the OR operation of two or more variables and then perform the AND operation of the result with a single variable, then the result will be similar to performing the AND operation of that single variable with each two or more variable and then perform the OR operation of that product. This law explains the process of factoring.

For three variables, the distributive law is written as:

$A(B + C) = AB + AC$

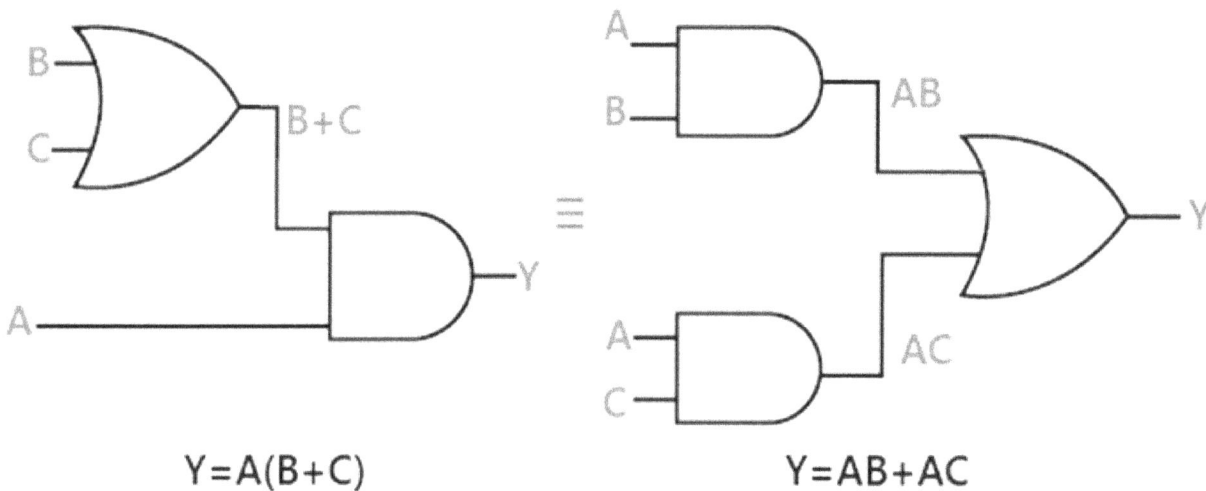

Y=A(B+C)   Y=AB+AC

Rules of Boolean algebra

There are the following rules of Boolean algebra, which are mostly used in manipulating and simplifying Boolean expressions. These rules plays an important role in simplifying boolean expressions.

1. A+0=A   7. A.A=A
2. A+1=1   8. A.A'=0
3. A.0=0   9. A"=A
4. A.1=A   10. A+AB=A
5. A+A=A   11. A+A'B=A+B
6. A+A'=1 12. (A+B)(A+C)=A+BC

Rule 1: A + 0 = A

Let's suppose; we have an input variable A whose value is either 0 or 1. When we perform OR operation with 0, the result will be the same as the input variable. So, if the variable value is 1, then the result will be 1, and if the variable value is 0, then the result will be 0. Diagrammatically, this rule can be defined as:

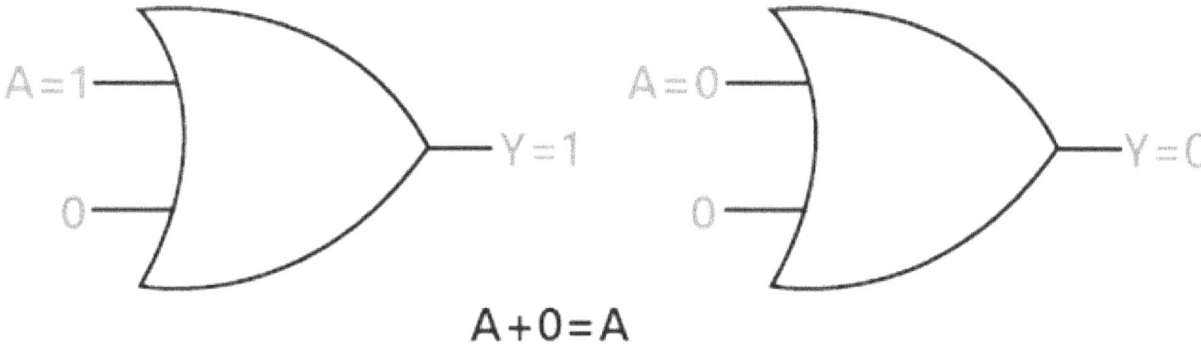

$$A+0=A$$

Rule 2: (A + 1) = 1

Let's suppose; we have an input variable A whose value is either 0 or 1. When we perform OR operation with 1, the result will always be 1. So, if the variable value is either 1 or 0, then the result will always be 1. Diagrammatically, this rule can be defined as:

$$Y=A+1=1$$

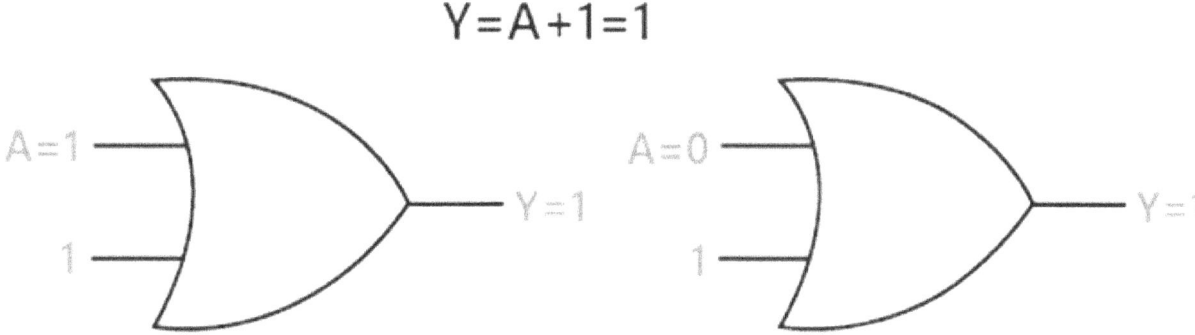

Rule 3: (A.0) = 0

Let's suppose; we have an input variable A whose value is either 0 or 1. When we perform the AND operation with 0, the result will always be 0. This rule states that an input variable ANDed with 0 is equal to 0 always. Diagrammatically, this rule can be defined as:

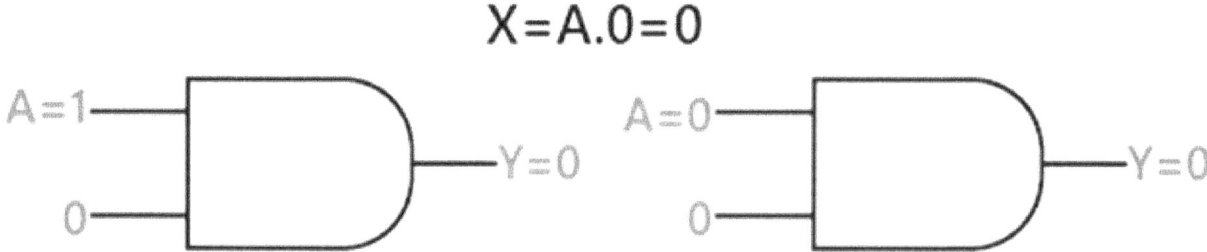

Rule 4: (A.1) = A

Let's suppose; we have an input variable A whose value is either 0 or 1. When we perform the AND operation with 1, the result will always be equal to the input variable. This rule states that an input variable ANDed with 1 is equal to the input variable always. Diagrammatically, this rule can be defined as:

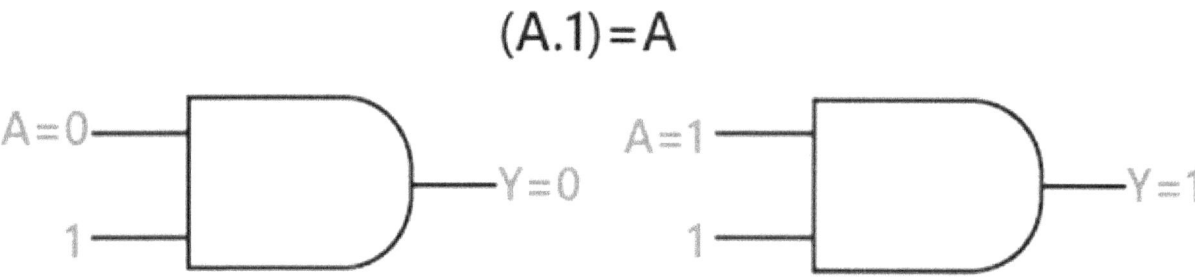

Rule 5: (A + A) = A

Let's suppose; we have an input variable A whose value is either 0 or 1. When we perform the OR operation with the same variable, the result will always be equal to the input variable. This rule states an input variable ORed with itself is equal to the input variable always. Diagrammatically, this rule can be defined as:

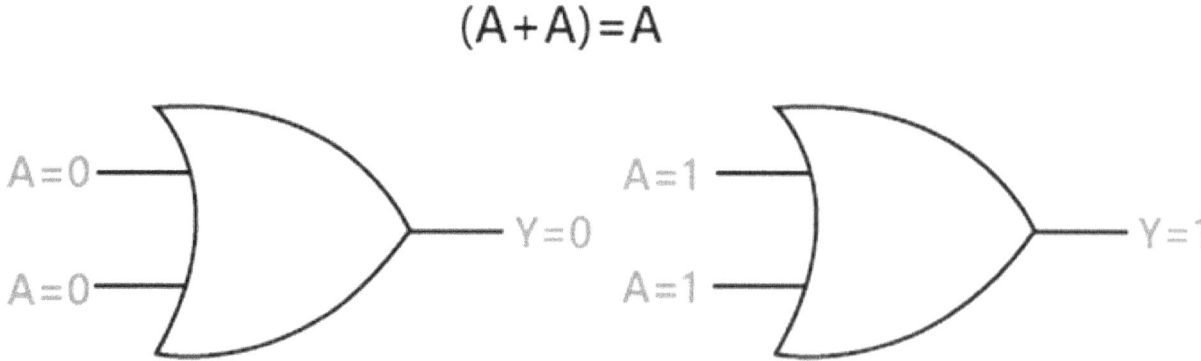

Rule 6: (A + A') = 1

Let's suppose; we have an input variable A whose value is either 0 or 1. When we perform the OR operation with the complement of that variable, the result will always be equal to 1. This rule states that a variable ORed with its complement is equal to 1 always. Diagrammatically, this rule can be defined as:

$$(A+A')=1$$

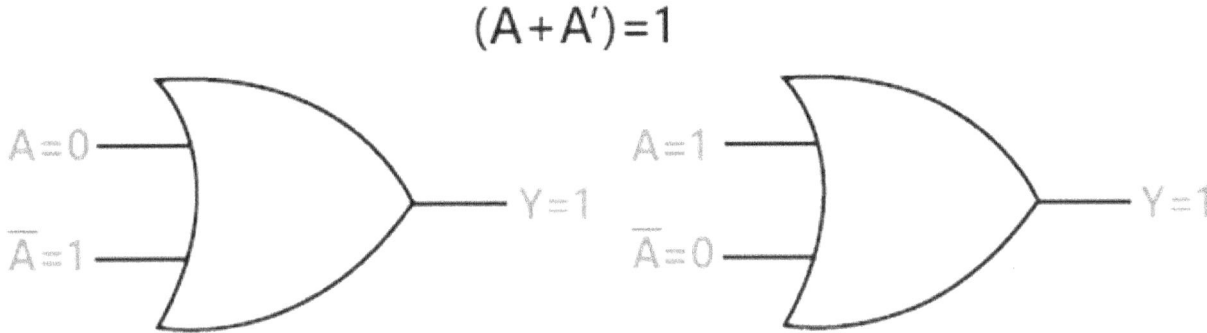

Rule 7: (A.A) = A

Let's suppose; we have an input variable A whose value is either 0 or 1. When we perform the AND operation with the same variable, the result will always be equal to that variable only. This rule states that a variable ANDed with itself is equal to the input variable always. Diagrammatically, this rule can be defined as:

$$(A.A)=A$$

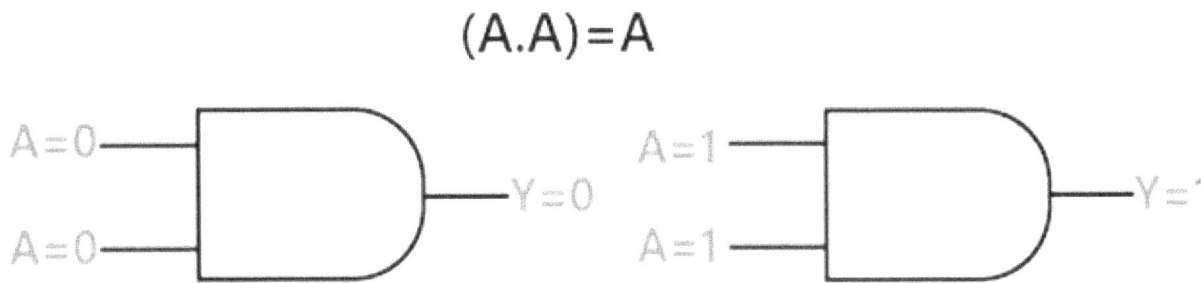

Rule 8: (A.A') = 0

Let's suppose; we have an input variable A whose value is either 0 or 1. When we perform the AND operation with the complement of that variable, the result will always be equal to 0. This rule states that a variable ANDed with its complement is equal to 0 always. Diagrammatically, this rule can be defined as:

$$(A.A') = 0$$

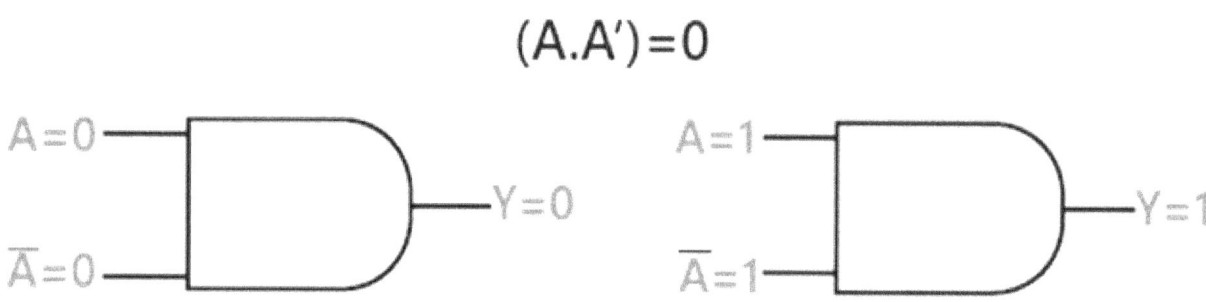

Rule 9: A = (A')'

This rule states that if we perform the double complement of the variable, the result will be the same as the original variable. So, when we perform the complement of variable A, then the result will be A'. Further if we again perform the complement of A', we will get A, that is the original variable.

$$A = (A')'$$

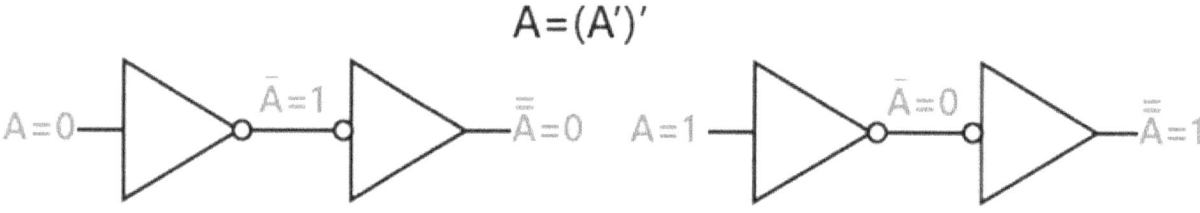

Rule 10: (A + AB) = A

We can prove this rule by using the rule 2, rule 4, and the distributive law as:

A + AB = A(1 + B)   Factoring (distributive law)

A + AB = A.1   Rule 2: (1 + B)= 1

A + AB = A   Rule 4: A .1 = A

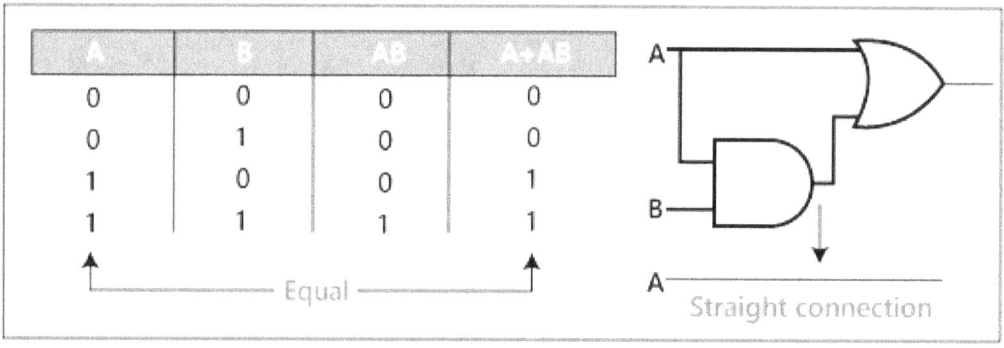

Rule 11: A + AB = A + B

We can prove this rule by using the above rules as:

A + AB = (A + AB)+ AB   Rule 10: A = A + AB

A+AB= (AA + AB)+ AB   Rule 7: A = AA

A+AB=AA +AB +AA +AB   Rule 8: adding AA = 0

A+AB= (A + A)(A + B)   Factoring

A+AB= 1.(A + B)   Rule 6: A + A = 1

A+AB=A + B   Rule 4: drop the 1

A	B	AB	A+AB	A+B
0	0	0	0	0
0	1	1	1	1
1	0	0	1	1
1	1	0	1	1

equal

Rule 12: (A + B)(A + C) = A + BC

We can prove this rule by using the above rules as:

(A + B)(A + C)= AA + AC + AB + BC   Distributive law

(A + B)(A + C)= A + AC + AB + BC   Rule 7: AA = A

(A + B)(A + C)= A(1 + C)+ AB + BC   Rule 2: 1 + C = 1

(A + B)(A + C)= A.1 + AB + BC   Factoring (distributive law)

(A + B)(A + C)= A(1 + B)+ BC   Rule 2: 1 + B = 1

(A + B)(A + C)= A.1 + BC   Rule 4: A .1 = A

(A + B)(A + C)= A + BC

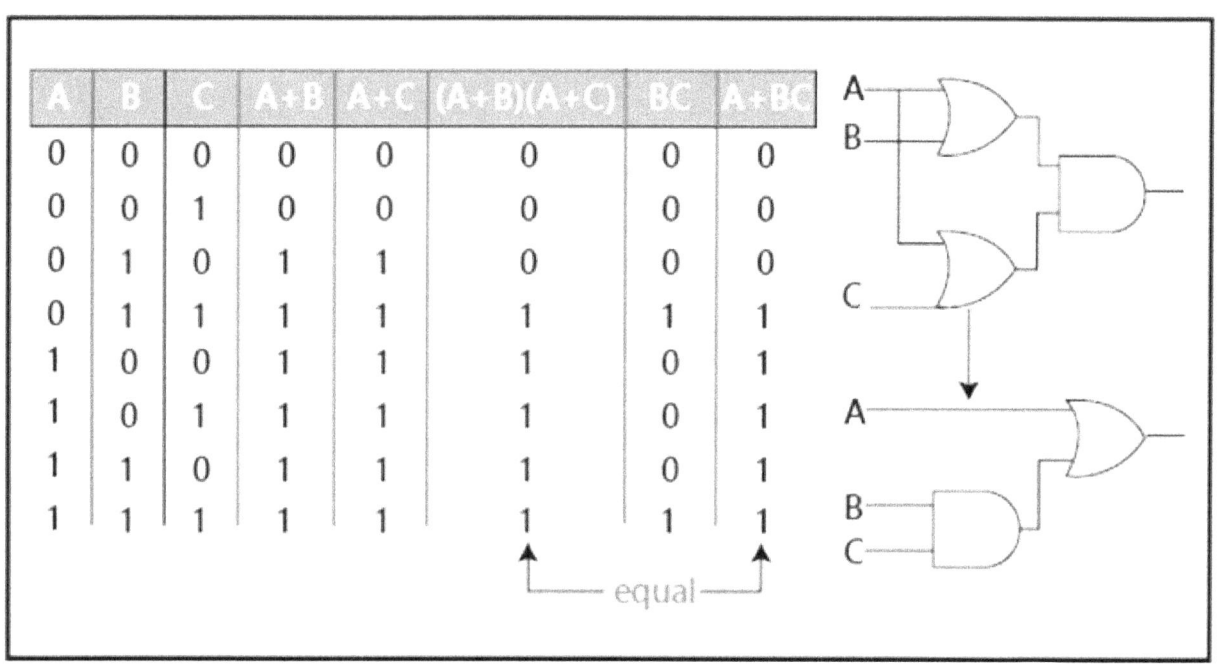

A	B	C	A+B	A+C	(A+B)(A+C)	BC	A+BC
0	0	0	0	0	0	0	0
0	0	1	0	0	0	0	0
0	1	0	1	1	0	0	0
0	1	1	1	1	1	1	1
1	0	0	1	1	1	0	1
1	0	1	1	1	1	0	1
1	1	0	1	1	1	0	1
1	1	1	1	1	1	1	1

equal

CHAPTER 4: LOGIC GATES

Logic gates play an important role in circuit design and digital systems. It is a building block of a digital system and an electronic circuit that always have only one output. These gates can have one input or more than one input, but most of the gates have two inputs. On the basis of the relationship between the input and the output, these gates are named as AND gate, OR gate, NOT gate, etc.

There are different types of gates which are as follows:

AND Gate

This gate works in the same way as the logical operator **"and"**. The AND gate is a circuit that performs the AND operation of the inputs. This gate has a minimum of 2 input values and an output value.

Y=A AND B AND C AND D……N
Y=A.B.C.D……N
Y=ABCD……N

Logic Design

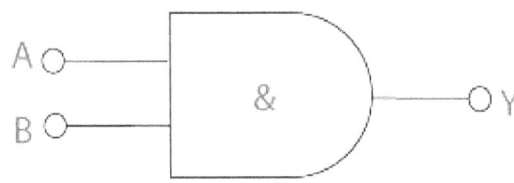

2- Input AND Gate

Truth Table

Inputs		Output
A	B	AB
0	0	0
0	1	0
1	0	0
1	1	1

OR Gate

This gate works in the same way as the logical operator **"or"**. The OR gate is a circuit which performs the OR operation of the inputs. This gate also has a minimum of 2 input values and an output value.

Y=A OR B OR C OR D……N
Y=A+B+C+D……N

Logic Design

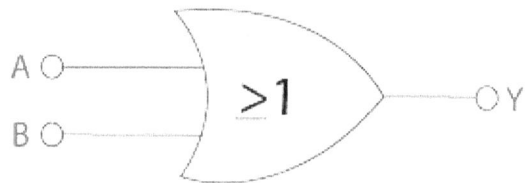

2- Input OR Gate

Truth Table

Inputs		Output
A	B	A+B
0	0	0
0	1	1
1	0	1
1	1	1

NOT Gate

The NOT gate is also called an inverter. This gate gives the inverse value of the input value as a result. This gate has only one input and one output value.

Y=NOT A
Y=A'

Logic Design

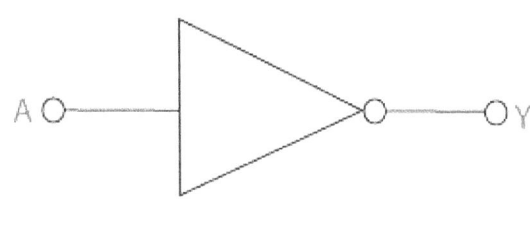

NOT Gate

Truth Table

Input	Output
A	B
0	1
1	0

NAND Gate

The NAND gate is the combination of AND gate and NOT gate. This gate gives the same result as a NOT-AND operation. This gate can have two or more than two input values and only one output value.

Y=A NOT AND B NOT AND C NOT AND D……N
Y=A NAND B NAND C NAND D……N

Logic Design

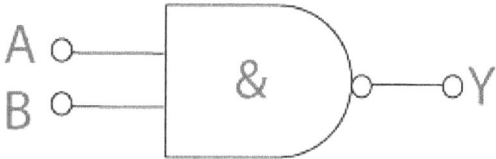

2- Input NAND Gate

Truth Table

Inputs		Output
A	B	(AB)'
0	0	1
0	1	1
1	0	1
1	1	0

NOR Gate

The NOR gate is the combination of an OR gate and NOT gate. This gate gives the same result as the NOT-OR operation. This gate can have two or more than two input values and only one output value.

Y=A NOT OR B NOT OR C NOT OR D……N
Y=A NOR B NOR C NOR D……N

Logic Design

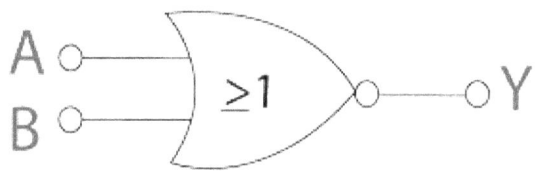

2- Input NOR Gate

Truth Table

Inputs		Output
A	B	(AB)'
0	0	1
0	1	0
1	0	0
1	1	0

XOR Gate

The XOR gate is also known as the Ex-OR gate. The XOR gate is used in half and full adder and subtractor. The exclusive-OR gate is sometimes called as EX-OR and X-OR gate. This gate can have two or more than two input values and only one output value.

Y=A XOR B XOR C XOR D......N
Y=A⊕B⊕C⊕D......N
Y=AB'+A'B

Logic Design

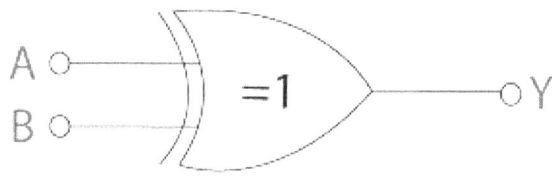

2-Input XOR Gate

Truth Table

Inputs		Output
A	B	A⊕B
0	0	0
0	1	1
1	0	1
1	1	0

XNOR Gate

The XNOR gate is also known as the Ex-NOR gate. The XNOR gate is used in half and full adder and subtractor. The exclusive-NOR gate is sometimes called as EX-NOR and X-NOR gate. This gate can have two or more than two input values and only one output value.

Y=A XNOR B XNOR C XNOR D……N
Y=A⊖B⊖C⊖D……N
Y=A'B'+AB

Logic Design

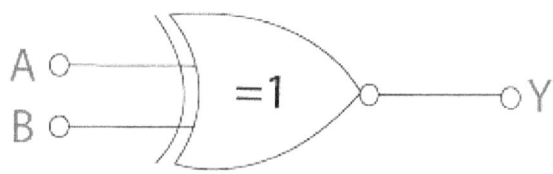

2-Input XNOR Gate

Truth Table

Inputs		Output
A	B	A⊖B
0	0	1
0	1	0
1	0	0
1	1	1

CHAPTER 5: AND GATE

The AND gate plays an important role in the digital logic circuit. The output state of the AND gate will always be low when any of the inputs states is low. Simply, if any input value in the AND gate is set to 0, then it will always return low output(0).

The logic or Boolean expression for the AND gate is the logical multiplication of inputs denoted by a full stop or a single dot as

A.B=Y

The value of Y will be true when both the inputs A and B are set to true.

Types of Digital Logic AND Gate

The AND gate is classified into three types based on the input it takes. These are the following types of AND gate:

The 2-input AND Gate

This is the simple formation of the AND gate. In this type of AND gate, there are only two input values and an output value. There are $2^2=4$ possible combinations of inputs. The truth table and logic design are given below:

Logic Design

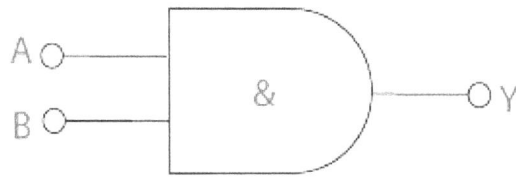

2- Input AND Gate

Truth Table

Input		Output
A	B	Y
0	0	0
0	1	0
1	0	0
1	1	1

The 3-input AND Gate

Unlike 2-input AND gate, the 3-input AND gate have three inputs. The Boolean expression of the logic AND gate is defined as the binary operation dot(.). The AND gate can be cascaded together to form any number of individual inputs. There are $2^3=8$ possible combinations of inputs. The truth table and logic design is given below:

Logic Design

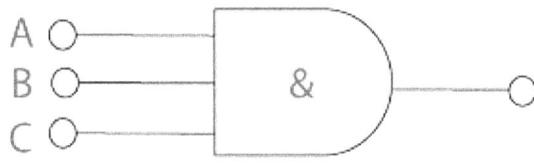

3- Input AND Gate

Truth Table

Input			Output
A	B	C	Y
0	0	0	0
0	0	1	0
0	1	0	0
0	1	1	0
1	0	0	0
1	0	1	0
1	1	0	0
1	1	1	1

The Multi-input AND Gate

In digital electronics, we can form n-input AND gate also. If there are n inputs, then (N/2)+1 AND gates will be used.

For example:

If we have 6 inputs A, B, C. D, E, F, then 4 AND gates are used in the logic design of 6-input AND gate. There is the following expression of the 6-input AND gate:

Y=(A.B).(C.D).(E.F)

In simple words, it is expressed as:

Y=A AND B AND C AND D AND E AND F

Logic Design

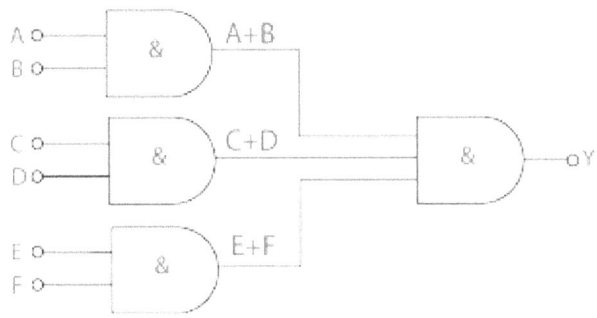

6- Input AND Gate

Truth Table

Input						Output
A	B	**C**	**D**	**E**	**F**	**Y**
0	0	0	0	0	0	0
0	0	0	0	0	1	0
0	0	0	0	1	0	0
0	0	0	0	1	1	0
.	.	.	.	.	.	.
.	.	.	.	.	.	.
.	.	.	.	.	.	.
1	1	1	1	0	0	0
1	1	1	1	0	1	0
1	1	1	1	1	0	0
1	1	1	1	1	1	1

CHAPTER 6: OR GATE

The OR gate is a mostly used digital logic circuit. The output state of the OR gate will always be low when both of the inputs states is low. Simply, if any input value in the OR gate is set to 1, then it will always return high-level output(1).

The logic or Boolean expression for the OR gate is the logical addition of inputs denoted by plus sign(+) as

A+B=Y

The value of Y will be true when one of the inputs is set to true.

Types of Digital Logic AND Gate

Just like AND gate, the OR gate is also classified into three types based on the input it takes. These are the following types of OR gate:

The 2-input OR gate

This is the simple form of the OR gate. In this type of OR gate, there are only two input values and an output value. There are $2^2=4$ possible combinations of inputs. The truth table and logic design are given below:

Logic Design

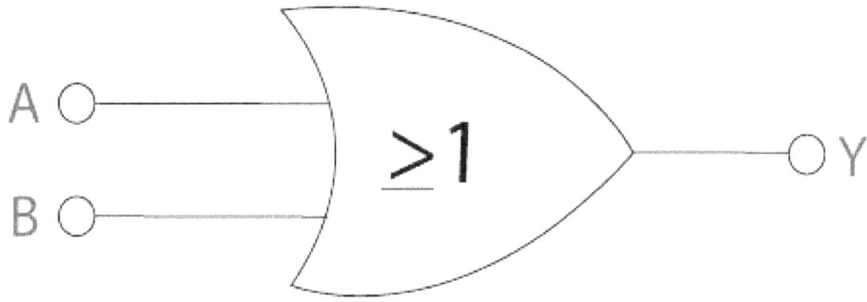

2- Input OR Gate

Truth Table

Input		Output
A	B	Y
0	0	0
0	1	1
1	0	1
1	1	1

The 3-input OR gate

Just like AND gate, the OR gate can also have any number of individual inputs. The Boolean expression of the logical OR gate is defined as the binary operation plus(+). Like AND gate, OR gate can also be cascaded together to form any number of individual inputs. There are $2^3=8$ possible combinations of inputs. The truth table and logic design are given below:

Logic Design

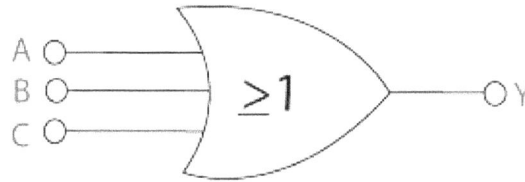

3- Input OR Gate

Truth Table

Input			Output
A	B	C	Y
0	0	0	0
0	0	1	1
0	1	0	1
0	1	1	1
1	0	0	1
1	0	1	1
1	1	0	1
1	1	1	1

The Multi-input OR Gate

The n-input OR gate can also be formed. If there are n inputs, then (N/2)+1 OR gates will be used.

For example:

If we have 6 inputs A, B, C. D, E, F, then 4 OR gates are used in the logic design of the 6-input OR gate. There is the following expression of the 6-input OR gate:

Y=(A+B)+(C+D)+(E+F)

In simple words, it is expressed as:

Y=A OR B OR C OR D OR E OR F

Logic Design

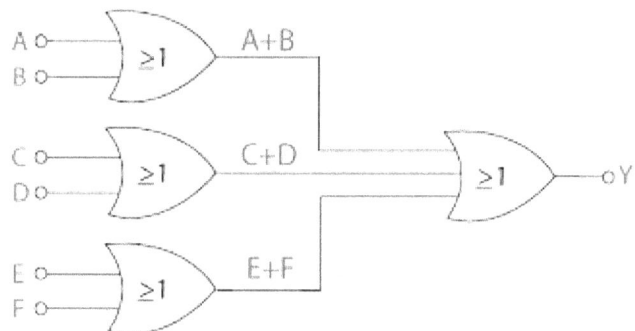

6- Input OR Gate

Truth Table

Input						Output
A	B	C	D	E	F	Y
0	0	0	0	0	0	0
0	0	0	0	0	1	1
0	0	0	0	1	0	1
0	0	0	0	1	1	1
.	.	.	.	.	.	.
.	.	.	.	.	.	.
.	.	.	.	.	.	.
1	1	1	1	0	0	1
1	1	1	1	0	1	1
1	1	1	1	1	0	1
1	1	1	1	1	1	1

CHAPTER 7: NOT GATE

The NOT gate is the most basic logic gate of all other logic gates. NOT gate is also known as an **inverter** or an **inverting Buffer**. NOT gate only has one input and one output. When the input signal is "Low", the output signal is "High" and when the input signal is "High", the output is "Low". The Boolean expression for the NOT gate is as follows:

A'=Y

When A is not true,then Y is true

The standard NOT gate is given a symbol that is shaped like a triangle with a circle at the end, pointing to the right. This circle is known as an "invert bubble" and is used to represent the logical operation of the NOT function in the NOT, NAND and NOR symbols in their output.

Logic Design

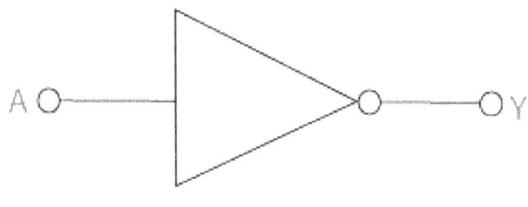

NOT Gate

Truth Table

Inputs	Outputs
A	Y
0	1
1	0

The complement value is generated by the NOT gate. The NOT gate is so-called because when the input signal is 0, the output signal will NOT be 0, Similarly, when the input signal is 1, the output signal will NOT be 1.

In the NOT gate, the bubble denotes the single inversion of the output signal. But this bubble can also exist on the gate's input to indicate an active-less input. This reversal of the input signal is not limited only to the NOT gate, but can also be used on any digital circuit or gate, as shown with the operation of inversion, whether it is at the input or output terminals. The easiest way is to think of the bubble as an inverter.

Use of Active-low Input Bubble

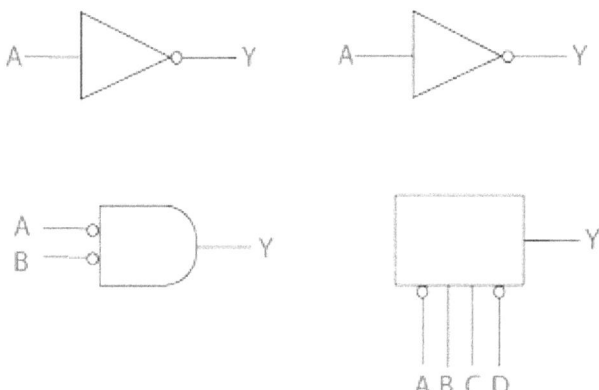

Bubble Notation for Input Inversion

Equivalent Gates

The NOT gate can also be formed with the help of the universal gates, i.e., NAND and NOR. For this, we have to connect both the inputs together to a common input signal. The NOT gate representation of NAND and NOR gate is as follows:

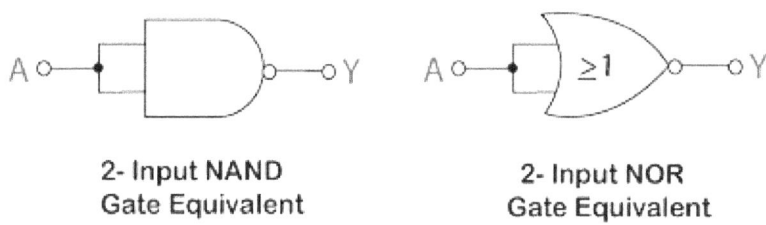

2- Input NAND
Gate Equivalent

2- Input NOR
Gate Equivalent

CHAPTER 8: NAND GATE

The NAND gate is a special type of logic gate in the digital logic circuit. The NAND gate is the universal gate. It means all the basic gates such as AND, OR, and NOT gate can be constructed using a NAND gate. The NAND gate is the combination of the NOT-AND gate. The output state of the NAND gate will be low only when all the inputs are high. Simply, this gate returns the complement result of the AND gate.

The logic or Boolean expression for the NAND gate is the complement of logical multiplication of inputs denoted by a full stop or a single dot as

$(A.B)'=Y$

The value of Y will be true when any one of the input is set to 0.

Types of Digital Logic AND Gate

The NAND gate is also classified into three types based on the input it takes. These are the following types of AND gate:

The 2-input NAND Gate

This is the simple formation of the NAND gate. In this type of NAND gate, there are only two input values and an output value. There are $2^2=4$ possible combinations of inputs. The truth table and logic design are given below:

Logic Design

2- Input "AND" gate plus a "NOT" gate

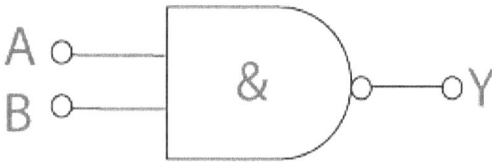

2- Input NAND Gate

Truth Table

Input		Output
A	B	Y
0	0	1
0	1	1
1	0	1
1	1	0

The 3-input NAND Gate

Unlike the 2-input NAND gate, the 3-input NAND gate has three inputs. The Boolean expression of the logic NAND gate is defined as the binary operation dot(.). The NAND gate can be cascaded together to form any number of individual inputs. There are $2^3=8$ possible combinations of inputs. The truth table and logic design are given below:

Logic Design

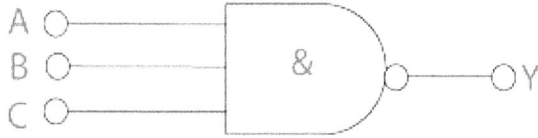

3- Input NAND Gate

Truth Table

Input			Output
A	B	C	Y
0	0	0	1
0	0	1	1
0	1	0	1
0	1	1	1
1	0	0	1
1	0	1	1
1	1	0	1
1	1	1	0

The Multi-input NAND Gate

Just like AND, NOT, and OR gate, we can also form n-input NAND gate. If the number of inputs required is odd, any "unused" input can be held high by directly connecting it to the power supply using high "suitable" pull-up resistors. There is the following expression of the 4-input NAND gate:

$Y=((A.B).(C.D))'$

In simple words, it is expressed as:

Y=A NAND B NAND C NAND D

Logic Design

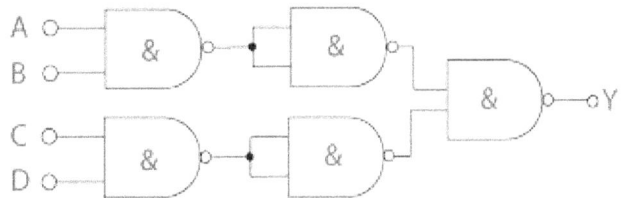

Truth Table

Input						Output
A	B	C	D	E	F	Y
0	0	0	0	0	0	1
0	0	0	0	0	1	1
0	0	0	0	1	0	1
0	0	0	0	1	1	1
.	.	.	.	.	.	.
.	.	.	.	.	.	.
.	.	.	.	.	.	.
1	1	1	1	0	0	1
1	1	1	1	0	1	1
1	1	1	1	1	0	1
1	1	1	1	1	1	0

AND gate using NAND gate

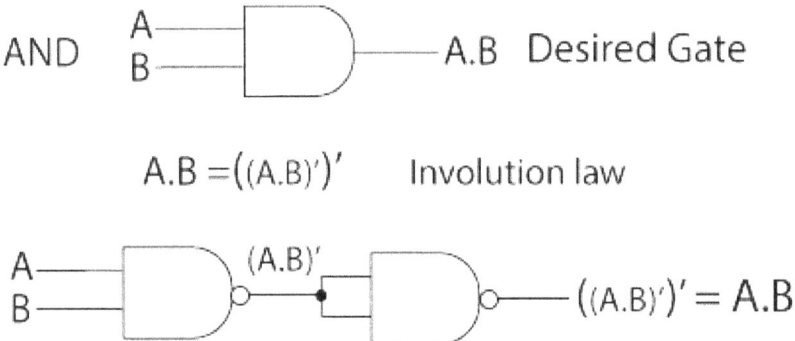

AND A B — A.B   Desired Gate

$$A.B = ((A.B)')'$$ Involution law

$$((A.B)')' = A.B$$

The Derivation of the AND Gate

OR gate using NAND gate.

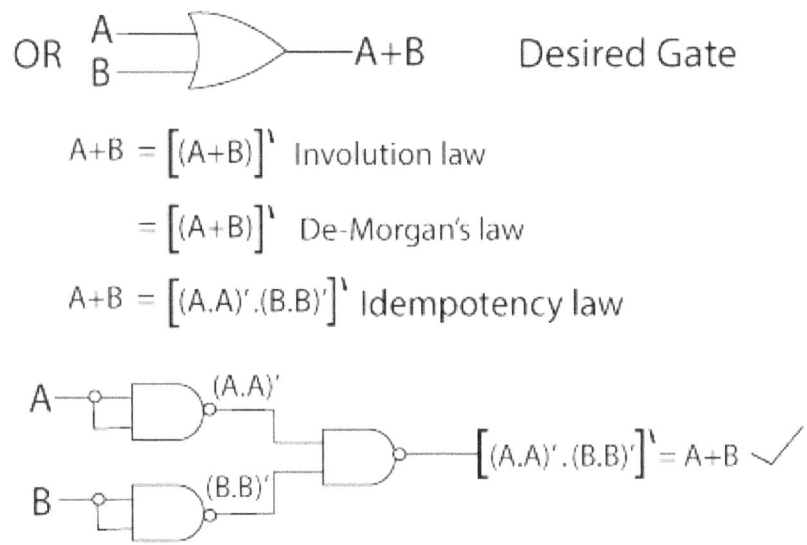

OR A B — A+B   Desired Gate

$$A+B = [(A+B)]'$$ Involution law

$$= [(A+B)]'$$ De-Morgan's law

$$A+B = [(A.A)'.(B.B)']'$$ Idempotency law

$$[(A.A)'.(B.B)']' = A+B \checkmark$$

The Derivation of the OR Gate

NOT gate using NAND gate

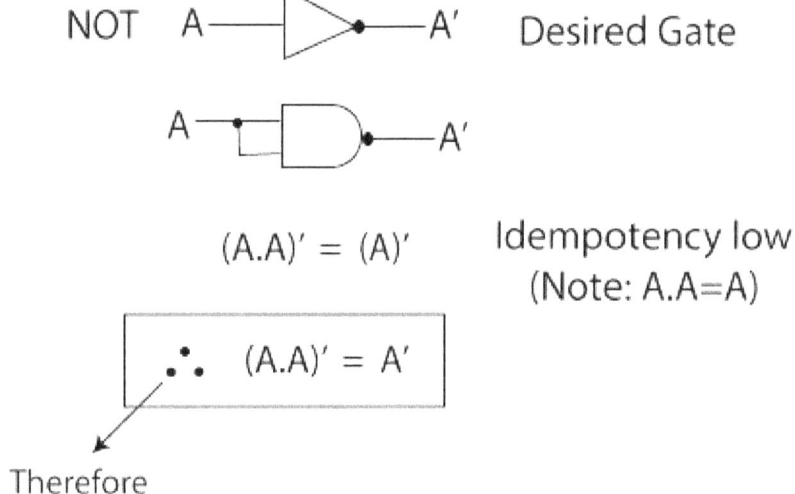

NOT $A \longrightarrow A'$ Desired Gate

$A \longrightarrow A'$

$(A.A)' = (A)'$ Idempotency low
(Note: A.A=A)

$\therefore \quad (A.A)' = A'$

Therefore

The Derivation of the NOT Gate

CHAPTER 9: NOR GATE

The NOR gate is also a universal gate. So, we can also form all the basic gates using the NOR gate. The NOR gate is the combination of the NOT-OR gate. The output state of the NOR gate will be high only when all of the inputs are low. Simply, this gate returns the complement result of the OR gate.

The logical or Boolean expression for the NOR gate is the complement of logical multiplication of inputs denoted by the plus sign as

$(A+B)'=Y$

The value of Y will be true when all of its inputs are set to 0.

Types of Digital Logic NOR Gate

The NOR gate is also classified into three types based on the input it takes. These are the following types of NOR gate:

The 2-input NOR gate

Just like other gates, it is also a simple form of the NOR gate. In this type of NOR gate, there are only two input values and an output value. There are $2^2=4$ possible combinations of inputs. The truth table and logic design are given below:

Logic Design

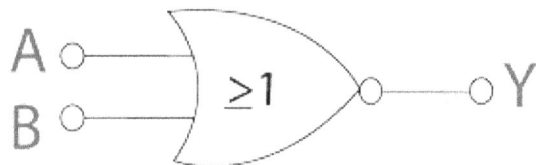

2- Input "AND" gate plus a "NOT" gate

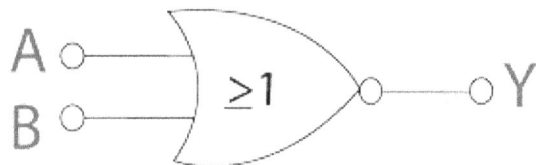

2- Input NOR Gate

Truth Table

Input		Output
A	**B**	**Y**
0	0	1
0	1	0
1	0	0
1	1	0

The 3-input NOR gate

Unlike the 2-input N gate, the 3-input NOR gate has three inputs. The Boolean expression of the logic NOR gate is defined as the binary operation addition (+). The NOR gate can be cascaded together to form any number of individual inputs. There are $2^3=8$ possible combinations of inputs. The truth table and logic design are given below:

Logic Design

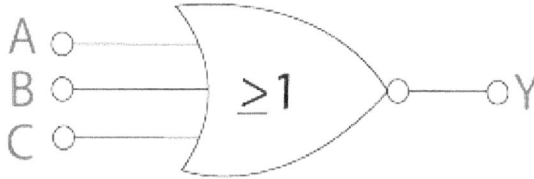

3- Input NOR Gate

Truth Table

Input			Output
A	B	C	Y
0	0	0	1
0	0	1	0
0	1	0	0
0	1	1	0
1	0	0	0
1	0	1	0
1	1	0	0
1	1	1	0

The Multi-input NOR Gate

Just like the NAND gate, we can also form the n-input NOR gate. If the number of inputs required is odd, any "unused" input can be held low by directly connecting it to the power supply using low "suitable" pull-up resistors. There is the following expression of the 4-input NOR gate:

Y=((A+B)+(C+D))'

In simple words, it is expressed as:

Y=A NOR B NOR C NOR D

Logic Design

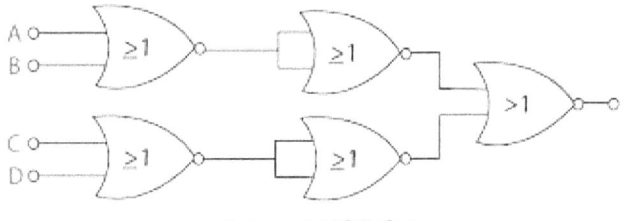

4- Input NOR Gate

Truth Table

Input						Output
A	B	C	D	E	F	Y
0	0	0	0	0	0	1
0	0	0	0	0	1	0
0	0	0	0	1	0	0
0	0	0	0	1	1	0
.	.	.	.	.	.	.
.	.	.	.	.	.	.
.	.	.	.	.	.	.
1	1	1	1	0	0	0
1	1	1	1	0	1	0
1	1	1	1	1	0	0
1	1	1	1	1	1	0

AND gate using NOR gate

AND Gate using NOR Gate

AND $\quad$ A— B— [AND gate symbol] —A.B $\quad$ Desired

$$A.B = \left[(A.B) \right]' \quad \text{Involution law}$$

$$= \left[A'+B \right]' \quad \text{Demorgan's law}$$

$$A.B = \left[(A+A)'+(B+B)' \right]' \quad \text{Idempotency law}$$

A — B — [NOR gate circuit] — $\left[(A+A)'+(B+B)' \right]' = A.B$

OR gate using NOR gate

OR Gate using NOR Gate

OR $\quad$ A— B— [OR gate symbol] — A+B $\quad$ Desired

$$A+B = \left((A.B)' \right)' \quad \text{Involution low}$$

A — B — [NOR gate circuit] $(A+B)'$ — $\left((A.B)' \right)' = A+B$

NOT gate using NOR gate

NOT Gate using NOR Gate

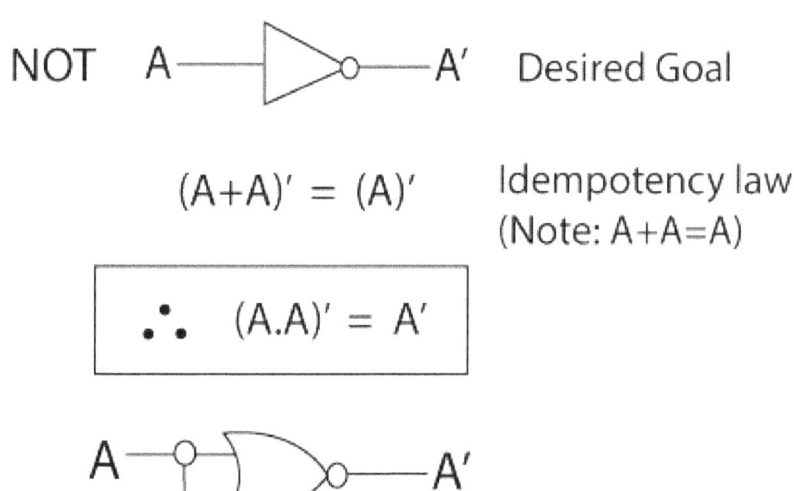

NOT   A —————▷o— A'   Desired Goal

$$(A+A)' = (A)'$$ Idempotency law
(Note: A+A=A)

$$\therefore \quad (A.A)' = A'$$

A —o⊃o— A'

CHAPTER 10: XOR GATE

The XOR gate stands for the Exclusive-OR gate. This gate is a special type of gate used in different types of computational circuits. Apart from the AND, OR, NOT, NAND, and NOR gate, there are two special gates, i.e., Ex-OR and Ex-NOR. These gates are not basic gates in their own and are constructed by combining with other logic gates. Their Boolean output function is significant enough to be considered as a complete logic gate. The XOR and XNOR gates are the hybrids gates.

The 2-input OR gate is also known as the Inclusive-OR gate because when both inputs A and B are set to 1, the output comes out 1(high). In the Ex-OR function, the logic output "1" is obtained only when either A="1" or B="1" but not both together at the same time. Simply, the output of the XOR gate is high(1) only when both the inputs are different from each other.

The plus(+) sign within the circle is used as the Boolean expression of the XOR gate. So, the symbol of the XOR gate is $\oplus$. This Ex-OR symbol also defines the "direct sum of sub-objects" expression. These are the following types of Exclusive-OR gate:

2-input Ex-OR gate

This is a simple form of the hybrid gate XOR. In this type of XOR gate, there are only two input values and an output value. There are $2^2=4$ possible combinations of inputs. The output level is high when both inputs are set to a different logic level. The Boolean expression of 2-input XOR gate is as follows:

$Y=(A \oplus B)$
$Y=(A' B + AB')$

The truth table and logic design are given below:

Logic Design

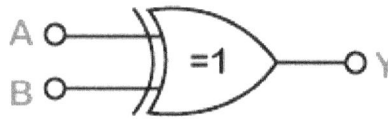

2-input "Ex-OR Gate

Truth Table

Input Output

A B Y

0 0 0

0 1 1

1 0 1

1 1 0

The 3-input XOR Gate

Unlike the 2-input XOR gate, the 3-input XOR gate has three inputs. There are $2^3=8$ possible combinations of inputs. The Boolean expression of the logical Ex-OR gate is as follows:

$Y=A\oplus B\oplus C$
$Y=A(BC)'+A' BC'+(AB)' C+ABC$

The truth table and logic design are given below:

Logic Design

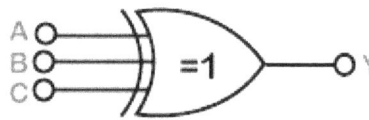

3-input Ex-OR Gate

Truth Table

Input			Output
A	B	C	Y
0	0	0	0
0	0	1	1
0	1	0	1
0	1	1	0
1	0	0	1
1	0	1	0
1	1	0	0
1	1	1	1

Ex-OR gate equivalent circuit

We can form the XOR or Ex-OR gate using gates such as AND, OR, and a universal gate NAND. The main disadvantage of this implementation is that we need to use different types of gates to form a single XOR gate. By using only the NAND gate, we can implement the Ex-OR gate also. This is an easier way of producing Ex-OR gate functionality.

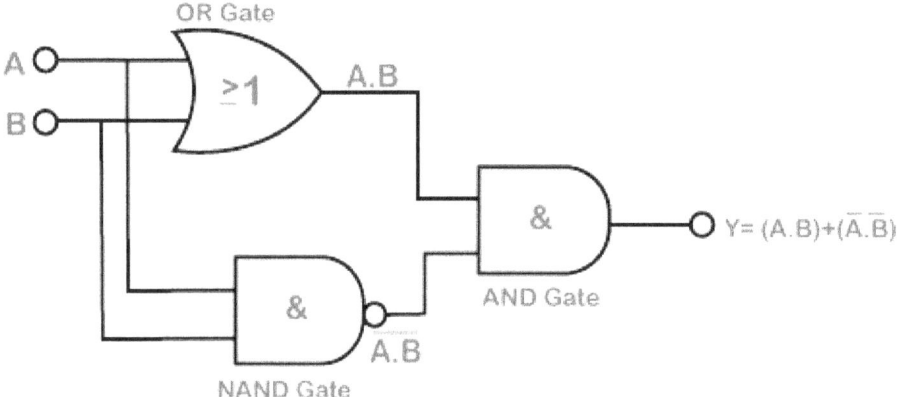

Ex-OR Function Realisation using NAND Gates

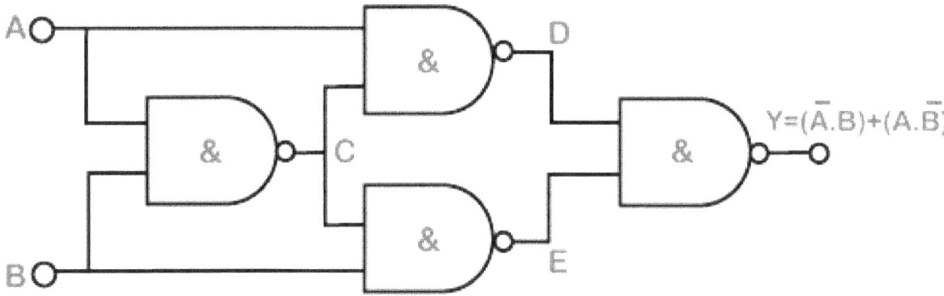

Use of Ex-OR gate

The Ex-or gate plays an important role in constructing digital circuits that perform arithmetic operations and calculations. Especially **Adders** and **Half-Adders**, as they can provide a "carry-bit" function or as a controlled inverter, where one input passes the binary data, and the other input is supplied with a control signal.

CHAPTER 11: XNOR GATE

The XNOR gate is the complement of the XOR gate. It is a hybrid gate. Simply, it is the combination of the XOR gate and NOT gate. The output level of the XNOR gate is high only when both of its inputs are the same, either 0 or 1. The symbol of the XNOR gate is the same as XOR, only complement sign is added. Sometimes, the XNOR gate is also called the **Equivalence gate**.

2-input Ex-NOR gate

It is a simple form of the hybrid gate XNOR. In this type of XNOR gate, there are only two input values and an output value. There are $2^2=4$ possible combinations of inputs. The output level is high when both inputs are set to high(1). The Boolean expression of 2-input XNOR gate is as follows:

$Y=(A{\oplus}B)'$
$Y=((AB)'+AB)$

The truth table and logic design are given below:

Logic Design

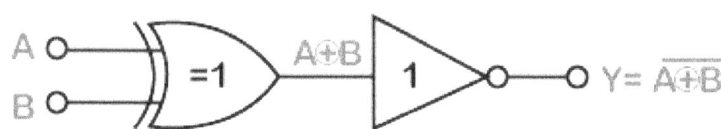

2-input "Ex-OR" Gate plus a " NOT" Gate

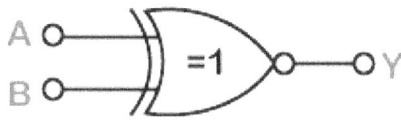

2-input "Ex-NOR Gate

Truth Table

Input Output

A	B	Y
0	0	1
0	1	0
1	0	0
1	1	1

The 3-input XNOR Gate

Unlike the 2-input XNOR gate, the 3-input XNOR gate has three inputs. There are $2^3=8$ possible combinations of inputs. The Boolean expression of the logical Ex-OR gate is as follows:

Y=(A⊕B⊕C)'
Y=(ABC)'+ABC'+AB'C+A'BC

The truth table and logic design are given below:

Logic Design

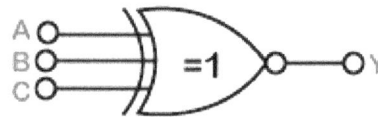

3-input Ex-NOR Gate

Truth Table

Input Output

A	B	C	Y
0	0	0	1

```
0 0 1 0
0 1 0 0
0 1 1 1
1 0 0 0
1 0 1 1
1 1 0 1
1 1 1 0
```

Ex-OR gate equivalent circuit

We can form the XNOR or Ex-NOR gate using the gates such as AND, OR, and NOT gate. The main disadvantage of this implementation is that we use different types of gates to form a single XNOR gate. By using the NAND gates only, we can implement the Ex-NOR gate also. This is an easier way of producing Ex-NOR gate functionality.

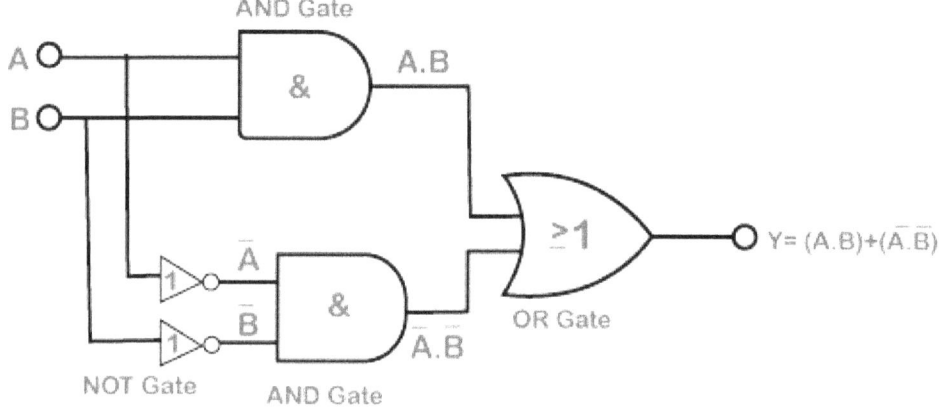

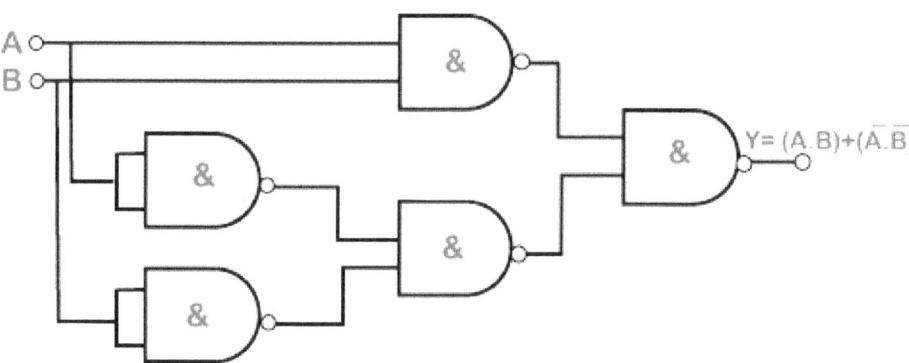

NAND Gate realisation

Use of Ex-OR gate

Ex-NOR gates are used mainly in electronic circuits that perform arithmetic operations and data checking such as *Adders*, *Subtractors* or *Parity Checkers*, etc. As the Ex-NOR gate gives an output of logic level "1," whenever its two inputs are equal, it can be used to compare the magnitude of two binary digits or numbers and so Ex-NOR gates are used in Digital Comparator circuits.